Mixtures and Solutions

by DANIEL GREENBERG

Table of Contents

Chapter 1
Mr. G's "No Problem" Hardware Store 2

Chapter 2
A Third (and Fourth) Type of Mixture 8

Chapter 3
Telling Them Apart 12

Chapter 4
More About Solutions 14

Chapter 5
Separating Mixtures 16

Chapter 6
Back to Mr. G's Store 18

Glossary 23

Index 24

Chapter 1

Mr. G's "No Problem" Hardware Store

Mr. G's hardware store has only one rule: NO PROBLEM. If you bought it here and it isn't right—return it. You'll get your money back.

Ms. Adams bought two pounds of steel nails. She took them home and mixed them with a half pound of brass screws. Now she wants her money back for the nails.

"No problem," says Mr. G.

Next comes Mr. Baez. He dissolved rock salt in a glass of water. Now *he* wants his money back, too.

"No problem," Mr. G says.

Mr. G knows that mixing and dissolving are examples of a **physical change**. The nails are still nails. The salt is still salt. Mixing or dissolving does not change any of their basic properties.

Mr. G says, "As long as you return the original materials, I'll give you your money back. No problem."

Physical Changes

- Changing shape
- Mixing
- Freezing or melting
- Cutting
- Stretching or bending
- Combining
- Dissolving

A piece of clay changing shape

A mixture (milk and chocolate syrup) being mixed

How can Ms. A and Mr. B get back the original materials they bought from Mr. G?

All Ms. A needs to do is separate the steel nails from the brass screws. To get the salt back, all Mr. B needs to do is boil off the water.

Steel nails mixed with brass screws

Clearly, Ms. A and Mr. B combined two substances together. But Ms. A made a **mixture**. Mr. B made a **solution**. How can you tell the difference?

A mixture happens when you combine any two (or more) substances together—steel nails and brass screws, oil and vinegar, nuts and bolts, rock salt and water. Making a mixture is an example of a physical change.

Physical changes alter the way something appears, not what it is. Changing shape is a physical change. Melting, freezing, or mixing are physical changes, too. The original ingredients are still there. They are just arranged in a different way.

Rock salt

Water

There are two types of mixtures. They are **heterogeneous** and **homogeneous** mixtures.

A simple heterogeneous mixture is chunky. Steel nails mixed with brass screws is an example of a heterogeneous mixture. Fruit salad, mixed nuts, and rocky road ice cream are, too. In each case, individual ingredients are chunky. You can pick them out.

A homogeneous mixture is not chunky. You cannot pick out individual ingredients. All solutions, such as Mr. B's rock salt and water, are homogeneous mixtures. In a solution, both things completely dissolve. Many solutions are clear when you look through them. They have no chunks or particles.

DON'T GET CONFUSED!

Is a solution a mixture?
Yes, it is a special type of mixture—a homogeneous mixture.

Heterogeneous Mixture

- chunky
- *heterogeneous* means "made up of different parts"
- uneven mixture
- can pick out ingredients
- not a solution

Fruit salad is a heterogeneous mixture.

Homogeneous Mixtures

- not chunky
- *homogeneous* means "made up of the same parts"
- even mixture
- cannot pick out ingredients
- all solutions are homogeneous

Clouds and cranberry juice are homogeneous mixtures.

To make a mixture of metals, you must turn the metals to liquid first. Very high temperatures melt metal.

Examples of solutions are seawater, air, and brass. In seawater, salt completely dissolves and disappears. You cannot see it, but you can taste it.

Air is a solution of gases. These gases are mostly nitrogen and oxygen. Brass is a solution of two solids—zinc and copper.

What do these solutions have in common? They are homogeneous mixtures. They are blended. They have no chunks. They are clear and smooth.

So far, we have seen two types of mixtures. The first is simple heterogeneous mixtures, such as fruit salad. The second is homogeneous solutions, such as salt and water.

Chapter 2

A Third (and Fourth) Type of Mixture

Mr. G is having a sale on toothpaste. Is toothpaste a heterogeneous or a homogeneous mixture?

Toothpaste is smooth and even like a solution—no chunks. But toothpaste has particles. They are just too small to see. So, what is toothpaste—a heterogeneous mixture or a solution?

Toothpaste is neither. It is a **colloid**. Colloids have properties that are somewhere between simple heterogeneous mixtures and solutions. They are smooth, but not too smooth.

Usually a colloid is a solid in a liquid. Yet colloids can be made of solids, liquids, or gases. Jellies, whipped cream, mayonnaise, rain clouds, and smoke are all colloids.

Colloids

- Jellies: solid fruit mixed into a liquid
- Mayonnaise: liquid oil and egg yolk mixed
- Whipped cream: a gas blended into liquid cream

Some colloids, like gels, appear to be see-through. Yet gels are not solutions. If you shine a flashlight beam through a colloid gel, you will see particles. Shining a beam through a true solution will show no such particles.

Flashlight Test
Shine a beam of light through a "clear" substance.
- If you see particles it is a colloid.
- If you don't see particles it is a true solution.

A fourth type of mixture is called a **suspension**. Muddy water, orange juice, as well as oil and vinegar salad dressing are suspensions.

Suspensions *look* like colloids. But if you wait, they will separate all on their own. For example, oil and vinegar salad dressing will stay mixed for a few minutes. Then they will separate.

A suspension of oil and vinegar will separate on its own.

Chapter 3

Telling Them Apart

We now have four different types of mixtures: simple heterogeneous mixtures, solutions, colloids, and suspensions. How do we tell them apart?

Mix something with water. Then follow the steps in the diagram. Do you see individual ingredients? Is the mixture clear or cloudy? Finally, shine a light on the mixture, or let it settle.

What kind of mixture do you have?

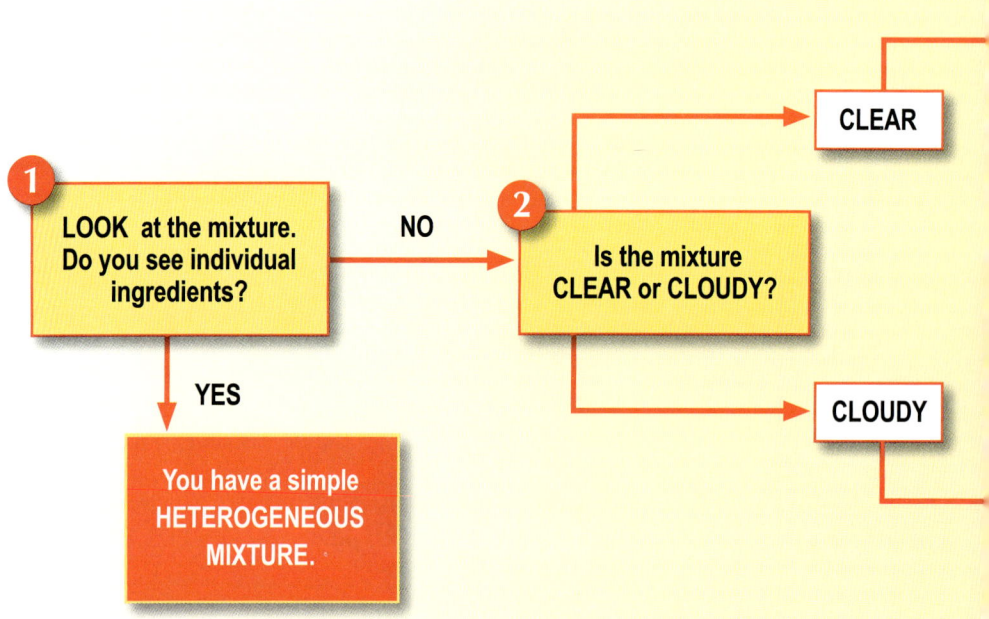

Follow these steps. You will be able to see if you have a simple heterogeneous mixture, a solution, a colloid, or a suspension.

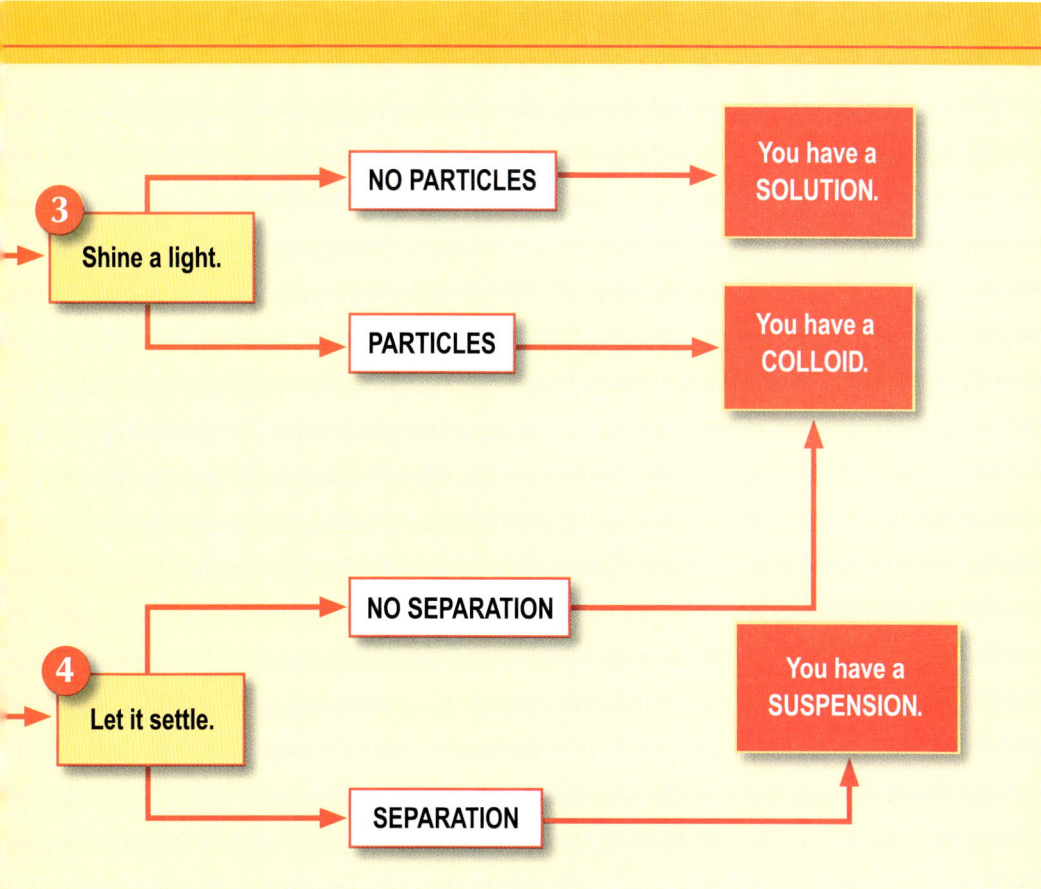

Chapter 4

More About Solutions

Like all mixtures, solutions can be made of solids, liquids, or gases. The ingredient that makes up most of a solution is called the **solvent**. The other ingredients are the **solutes**.

In sugar water, sugar is the solute. Water is the solvent. Nitrogen makes up 78 percent of air. Nitrogen is air's solvent. Oxygen and other gases are solutes.

Every solute and solvent pair are different. For example, table salt will dissolve in water. But if you add table salt to rubbing alcohol, it will not dissolve. Salt is not soluble in alcohol.

Temperature affects how much of a solute will dissolve. Add sugar to cold water. Soon, you will start to see leftover sugar in the bottom of the cup. If you heat the water, you can dissolve much more sugar.

More sugar will dissolve in heated water than in cold water.

Chapter 5

Separating Mixtures

Any physical change can be reversed. Water can freeze. Then it can melt again. A rubber band can stretch to a new shape. Then it can come back to its old shape. Making a mixture is an example of a physical change. Any mixture can be unmixed, or separated into its original parts.

Different types of mixtures are separated in different ways. Separating heterogeneous mixtures can be easy. Picking out steel nails from brass screws is no problem. Separating different ingredients in beach sand can be a bit harder. Yet it can be done.

Separating Saltwater

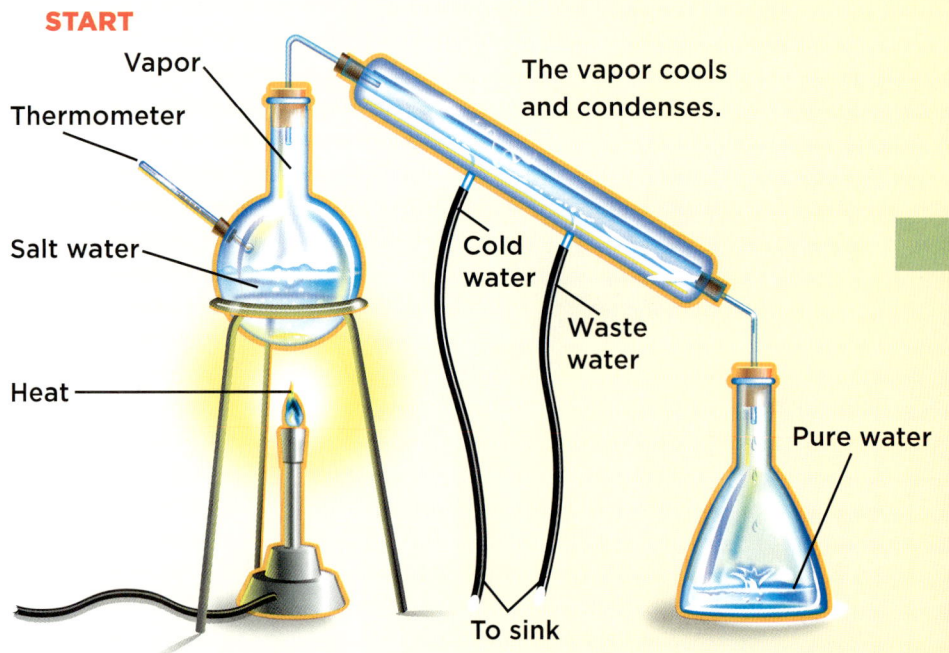

Separating a suspension often requires a filter. Pour muddy water through a filter. The larger soil particles get caught in the filter. The smaller water particles pass right through.

Separating a colloid or solution can be harder. To separate salt from seawater, heat the solution until it boils. Collect all the water that boils away. You will be left with only the salt. This method for separating mixtures is called distillation.

Chapter 6

Back to Mr. G's Store

Mr. G will take back *anything*, as long as it is the same thing he sold you. Mr. G had no problem taking back steel nails that were mixed with brass screws. He had no trouble taking back salt that was mixed with water.

But what about an iron bucket that has turned to rust? Will Mr. G take it back?

"No I won't," says Mr. G. "The iron went through a **chemical change**. It is no longer iron. It is rust."

Chemical changes are very different from physical changes. In a physical change, substances get rearranged: mixed, re-shaped, heated, or frozen.

In a chemical change, things are not re-shaped or mixed. *New* substances are made. For example, the bucket that Mr. G's customer wanted to return was rusty. Iron combined with oxygen turns into a new substance—rust.

How is rust different from iron? Iron is strong. Rust is weak. Iron is silver-colored; rust is dull red. Clearly, rust is a new substance.

A Chemical Change—Rusting

Iron or steel
- strong
- silver color
- strongly attracted by magnet
- good electrical conductor

Rust
- weak
- red color
- weakly attracted by magnet
- poor electrical conductor

The iron in this rusty bucket has gone through a chemical change.

What are some other chemical changes? Fire turns wood into ash and smoke. Plants turn carbon dioxide and water into sugars. Electricity turns water into oxygen and hydrogen gas.

In each case, the ingredients you ended up with were different from the ones with which you started. For example, when hydrogen gas and oxygen gas come together, they form water. Yet hydrogen and oxygen are *nothing* like water. They are not wet. They are not liquid. They are colorless gases.

When chemical changes happen, a new substance is formed. Cement, sand, and water go through a chemical change to make concrete. They cannot be separated. Ash cannot go back to being wood again.

Ice is frozen water. When the temperature goes above freezing, ice will return to its liquid state.

So that's why Mr. G will not give back the money for the bucket. Some of the original iron is gone. It has been replaced by rust, a new substance.

"When it's a physical change—no problem," Mr. G says. "But a chemical change—that's a different thing."

Glossary

chemical change (KEM-i-kuhl CHAYNJ) a change in a substance that happens when atoms link together in a new way, making a new substance *(page 18)*

colloid (KOL-oyd) a thick substance formed when very small particles that cannot be dissolved stay scattered throughout a liquid, solid, or gas without sinking *(page 8)*

heterogeneous (het-uhr-uh-JEE-nee-uhs) made up of different parts; not the same *(page 6)*

homogeneous (hoh-muh-JEE-nee-uhs) made up of the same parts *(page 6)*

mixture (MIKS-chuhr) a combination of substances that are blended together without forming new substances *(page 5)*

physical change (FIZ-i-kuhl CHAYNJ) a change of matter in size, shape, or state without forming a new substance *(page 3)*

solute (SOL-yewt) a substance that is dissolved by another substance to form a solution *(page 14)*

solution (suh-LEW-shuhn) a mixture of substances blended so it looks the same everywhere *(page 5)*

solvent (SOL-vuhnt) a substance that dissolves one or more substances to form a solution *(page 14)*

suspension (suh-SPEN-shuhn) a mixture in which suspended particles can be seen *(page 11)*

Index

chemical change, 18–22

colloid, 8–13, 17

combining substances, 5–7, 14, 19–20

distillation, 16–17

gases, 7, 9, 14, 20

gels, 10

homogeneous mixtures, 6–8

liquids, 7, 9, 14, 20–21

mixtures, 3, 5–8, 11–14, 16–17

particles, 6, 8, 10, 13

physical change, 3, 5, 16, 18, 22

separating mixtures, 11, 13, 16–17

simple heterogeneous mixtures, 6–8, 12–13

solids, 7–9, 14

solute, 14–15

solution, 5–8, 10, 12–14, 17

solvent, 14–15

suspension, 11–13, 17

temperature, 14, 21